Science Vocabulary

force
A **force** is a push or a pull.

These bikers use **force** to move.

motion
When an object is moving, it is in **motion.**

These wheelchair racers are in **motion.**

gravity

Gravity is a force that pulls things toward the center of Earth.

Gravity pulls these skaters down the hill.

magnet

A **magnet** is an object able to pull some metals toward itself.

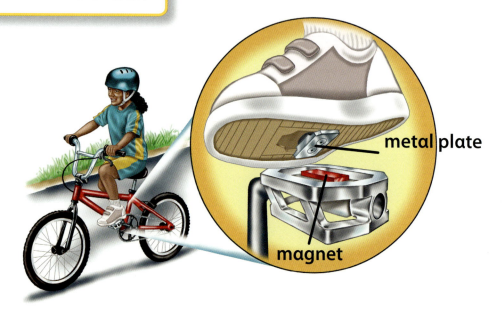

metal plate

magnet

This bicycle pedal has a **magnet** that pulls on a metal piece in the shoe.

pole

A **pole** is the part of a magnet where its force is the strongest.

The force of the magnet is strongest at its north **pole** and south **pole.**

attract

To **attract** is to pull toward.

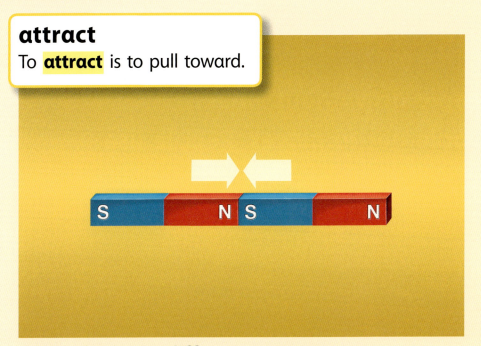

Poles that are different **attract,** or pull together.

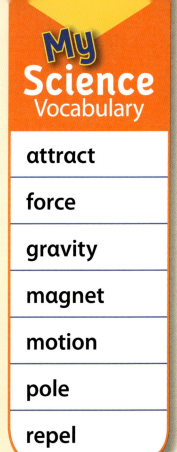

My Science Vocabulary

attract

force

gravity

magnet

motion

pole

repel

repel

To **repel** is to push away.

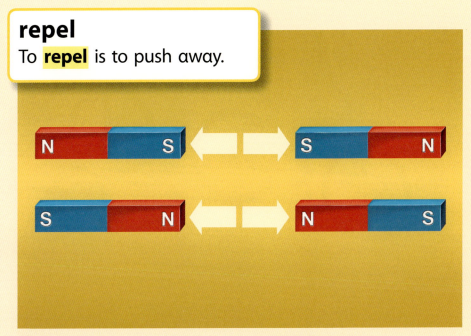

Poles that are the same **repel,** or push away.

On Wheels

What would people do without wheels? Wheels help people get to places. People use wheels for many different sports.

Bike riders use **force** to pedal and make the bike wheels turn. A force is a push or a pull. Riders push and pull to move their handlebars and their wheels.

force

A **force** is a push or a pull.

Forces and Motion

People race down the road. They use their arms to push and pull the wheels. Force puts these wheelchairs in **motion**.

The wheelchairs are rolling. They are in motion.

motion

When an object is moving, it is in **motion**.

This person uses force to stop a wheelchair. The person pulls on the wheel to slow it down. The wheel pushes against the ground to come to a stop.

This person has fun skating on a sidewalk. She gives small pushes with her feet. Small pushes help her roll along slowly.

This skater uses a lot of force to race. She pushes hard off the ground to make her wheels turn faster. She uses force to move quickly.

Look at these pictures. People are using wheels in different ways. They use force to put wheels in motion. Some people use pushes. Others use pulls.

Bike

Stroller

Scooter

Wagon

How do these pictures show pushes and pulls?

Push	Pull

push

pull

push

pull

A biker needs force to pedal a bicycle and to make it stop. To stop the bicycle, the biker presses on the brakes. This makes the brake pads press on the wheels.

brake pads

The wheels slow down and stop. The force of the wheel pushing against the ground stops the bike.

A biker uses force to go up and down a bridge on a back road trail.

The bridge sends the biker into the air. The biker uses the handlebars to control his motion and direction.

Gravity

A girl on a skateboard rolls downhill because of **gravity**. Earth's gravity is a force that pulls down on everything all the time.

gravity
Gravity is a force that pulls things toward the center of Earth.

This skateboarder is soaring through the air.
Gravity will bring him down to the ground.

Gravity keeps these skaters on the ground. It also pulls them down the hill.

Gravity makes it harder to go uphill. People have to use more force to push against gravity. These people would need less force to move on a flat road.

Magnets

Magnets make a pulling force. They attract, or pull, some metals such as iron.

magnet

A **magnet** is an object able to pull some metals toward itself.

attract

To **attract** is to pull toward.

Some bike pedals are magnetic. The magnets help bikers keep their feet on their pedals as they ride on the course.

Magnets on the bike pedals attract the metal plates in the biker's shoes. If the biker falls, the magnet pulls apart from the metal in her shoes.

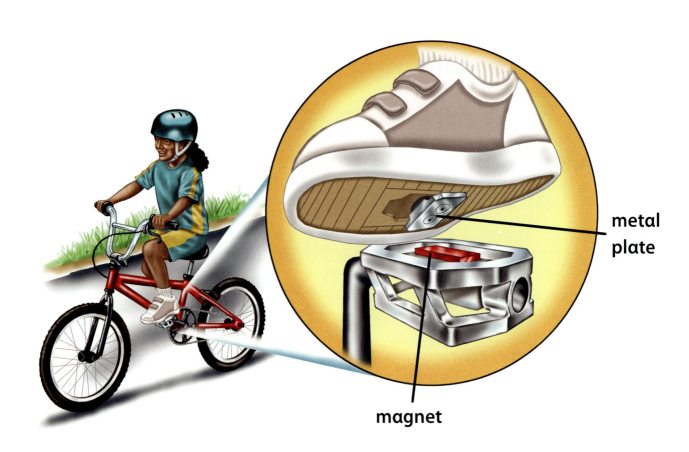

metal
plate

magnet

Magnets have two **poles.** Magnets have north and south poles. Poles that are different pull together. Poles that are the same, **repel,** or push away.

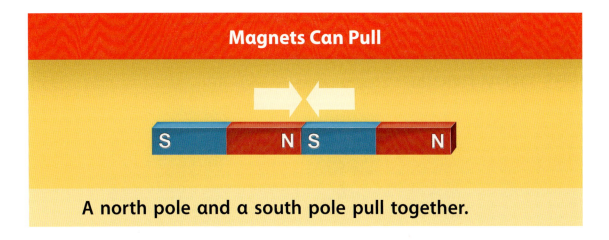

Magnets Can Pull

A north pole and a south pole pull together.

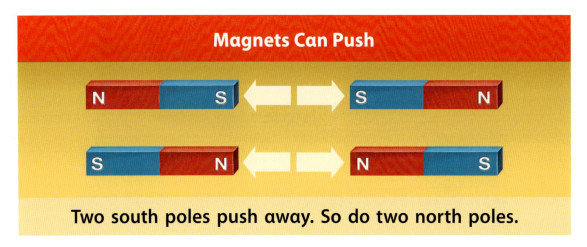

Magnets Can Push

Two south poles push away. So do two north poles.

pole

A **pole** is the part of a magnet where its force is the strongest.

repel

To **repel** is to push away.

Conclusion

People use forces to move on wheels. They use more force to go up a hill. They use less force to come down a hill. Gravity pulls them back down. Pushes and pulls help bikers move fast. They also help bikers stop. Forces help people get places on wheels.

Think About the Big Ideas

1. How are forces used with wheels?
2. Why is gravity important in skating or skateboarding?
3. How are magnets used on bikes?

Share and Compare

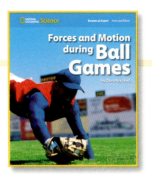

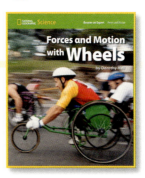

Turn and Talk

Look at the pictures of magnets in your books. Compare the different ways magnets are used.

Read

Find your favorite part of the book and read it to a partner.

Write

Tell about forces in your book. Share what you wrote with a classmate.

Draw

Draw an example of a force. Share your drawing with a classmate.

Meet Marianne Dyson

Marianne Dyson worked for NASA. She helped astronauts plan their days in space. During a space shuttle flight, parts of the shuttle stopped working. The flight had to be cut from five days to two days.

The crew could not do all the tests planned for the flight. To solve the problem, Marianne made a list of the most important tests. The crew used the new plan to finish these tests. Marianne's plan helped make the flight a success.